IMAGINATION VACATION
COLORADO

Written and Illustrated by Anastasia Kierst
In consultation with Christopher Kierst, geologist

Imagination Vacation Colorado:
Volume 2 in the Imagination Vacation Series
Published by Eternal Summers Press LLC
eternalsummerspress.com

Copyright © 2014 Anastasia Kierst
Edited by Ashley Argyle
First Edition
Published 2014
Second Printing, July 2014
Printed in PRC

All rights reserved.
No part of this book, aside from the four educator resource pages, may be reproduced by any means, except for purposes of review, without permission in writing from the author.
ISBN-10:0-9896337-3-X
ISBN-13:978-0-9896337-3-4

TO MY SISTER LEXI

Thanks for making it all so much more interesting! We've had some crazy adventures and I'm so glad I had you to share them.

"The joy of youth is discovery."
- Jeff Martin

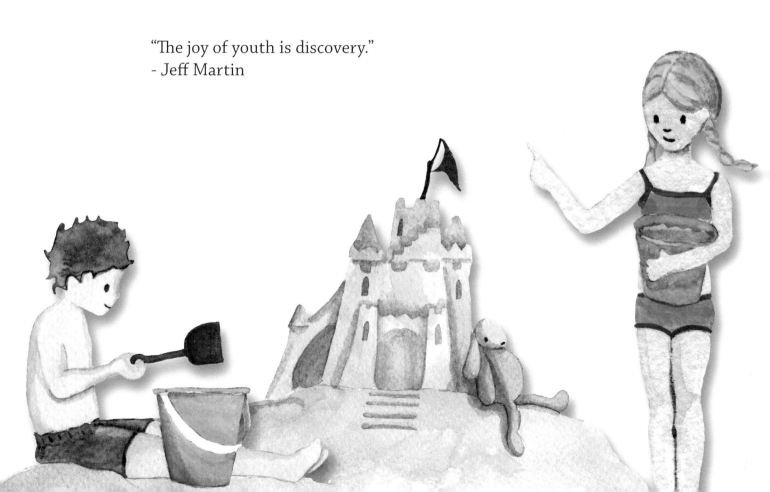

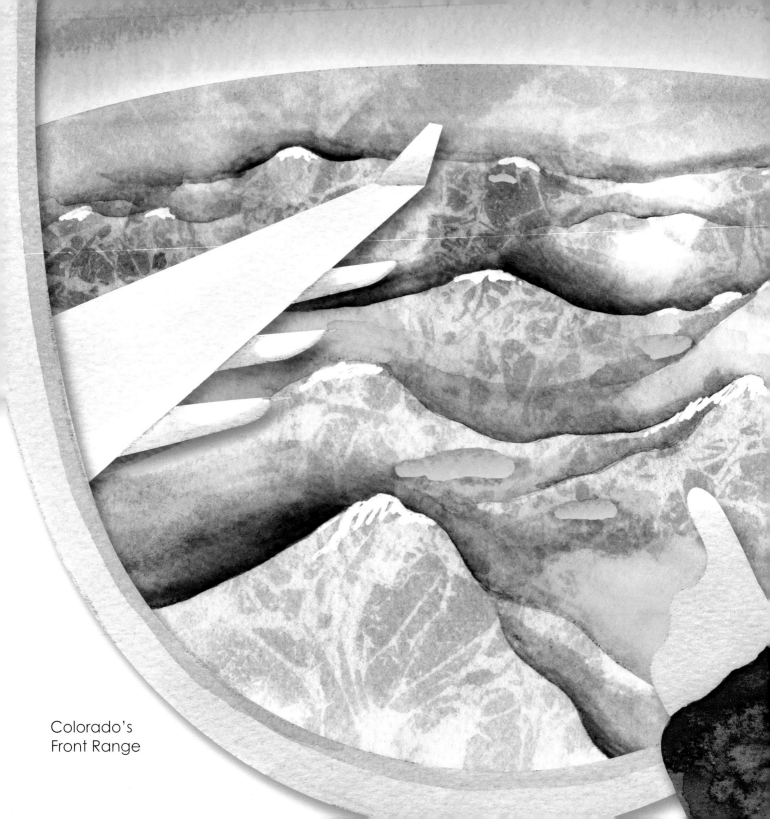

Colorado's Front Range

"Mom?"
"What are you imagining, Emmaline?"
"I think those little blue dots are...

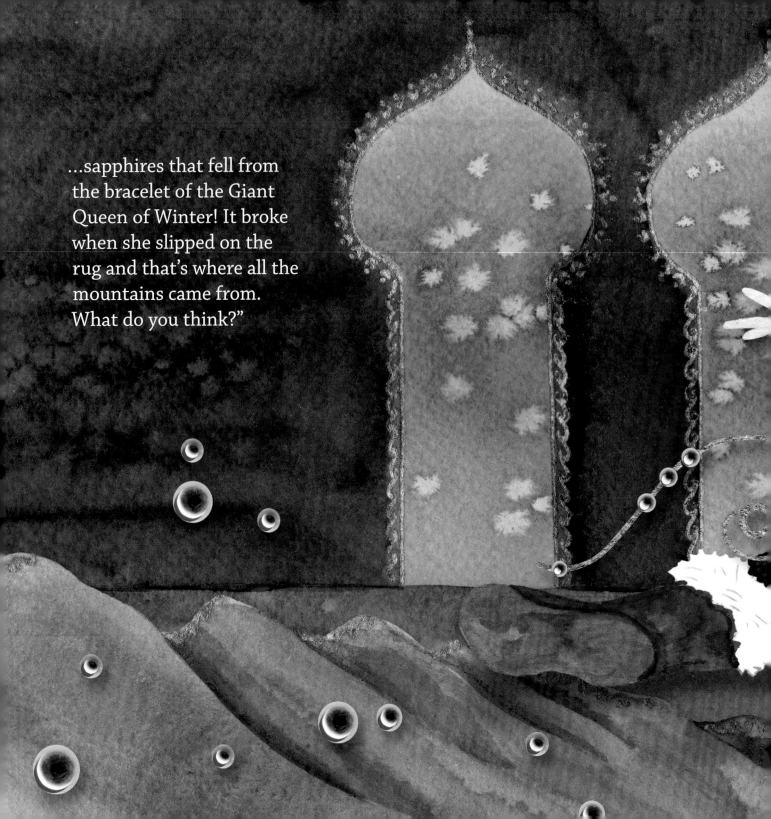

...sapphires that fell from the bracelet of the Giant Queen of Winter! It broke when she slipped on the rug and that's where all the mountains came from. What do you think?"

"Well, mountains are formed in several ways. Pikes Peak and the other mountains you see down there were formed by *plate tectonics*. The earth is made up of layers. The outside layer, where we live, is called the *crust*. It is part of the *lithosphere*, which is a hard, rocky layer.

The lithosphere is made up of 52 separate pieces called plates.

Each plate rests on the *mantle*, which is made up of hot, partly melted rock. The plates can slide around kind of like the rug on the floor you imagined. When the plates run into each other, one has to slide under the other.

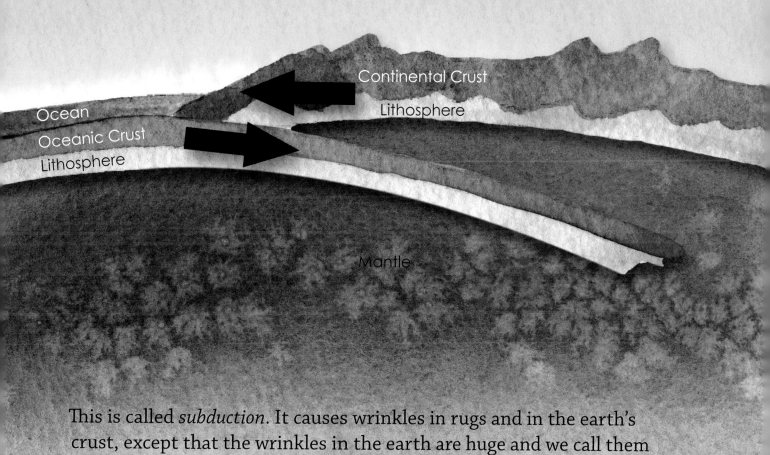

This is called *subduction*. It causes wrinkles in rugs and in the earth's crust, except that the wrinkles in the earth are huge and we call them mountains! As for the little blue dots, let's save that for another day."

Loch Vale, Rocky Mountain National Park

"Last winter about a thousand kids climbed up there and built a giant snow fort. It was so big and they used so much snow that it lasted all summer! What do you think, Dad?"

"Well there is a lot of snow up there, Oliver. But your snow fort is actually a *glacier*, a huge, heavy lump of ice that forms when snow piles up faster than it melts. A glacier has so much ice that it doesn't melt away during the summer. Rocky Mountain National Park was covered by them in the last ice age.

lateral moraine

wooly mammoth

muskox

horn

glacier

smilodon

As the glaciers slid slowly down from the mountains, they shaped the land in interesting ways. They pushed rocks into piles called *moraines* and even carved out big u-shaped valleys. Emmaline, remember the sapphires you noticed from the plane? Those are actually lakes left behind when the glaciers melted."

"Mom, Dad! I know where all this sand came from!"
"What are you thinking, Emmaline?"

Great Sand Dunes National Park

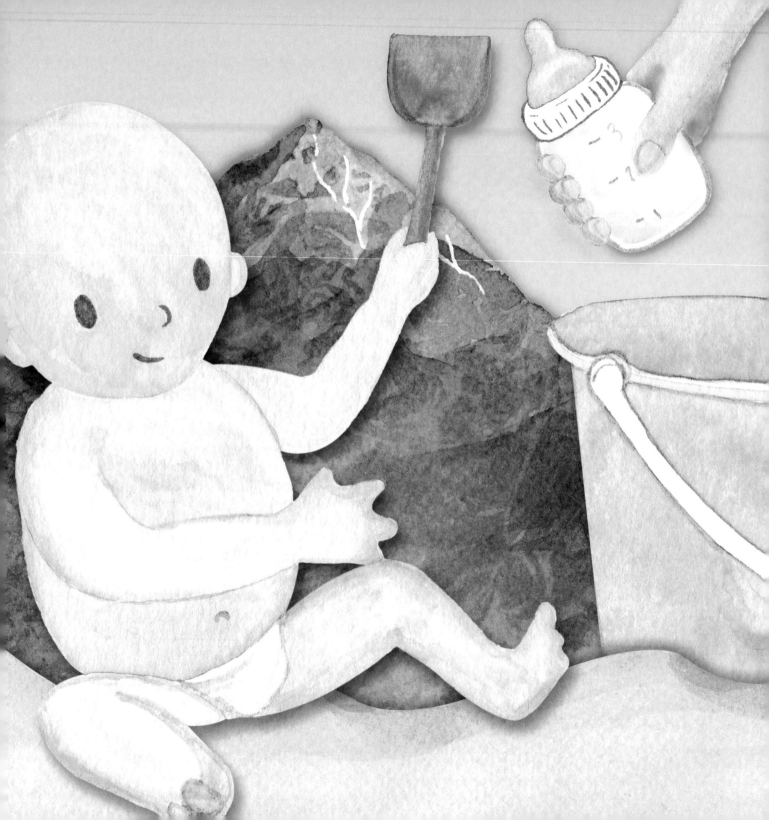

"The giants' baby was getting into trouble so they built him a sandbox to play in!"
"That wasn't it!" Oliver interrupted.
"What was it really, Mom?"

"In this case, there is a giant playing with the sand – the wind! Sand is carried to the valley by rivers and streams. The wind sweeps that sand off the valley floor and carries it through the air.

sand dunes

waterborne sand

When the wind hits the Sangre de Cristo mountains, it slows down and the sand lands here, in Great Sand Dunes National Park. The wind keeps pushing the sand up against the mountains and the streams keep carrying the sand back to the valley."

"Wow! I know what this is!"
"What is it, Oliver?" Mom asked.

"It's a dinosaur sandwich! T-rex got hungry, so he made a dino sandwich. But before he could eat it, a volcano erupted. He ran away and his sandwich was covered with rocks. That is what you see here!"
"Seriously? A dino sandwich? DAD!" Emmaline groaned.

"Well, it was kind of like a sandwich. These dinosaurs are *fossilized*.

When the dinosaurs died, they were covered by *sediments*: small rocks like gravel, sand and silt. The sediments formed layers, like a sandwich.

stegosaurus

sediments

Over time, water in the ground soaked into their bones and the minerals carried in that water replaced the bones, creating a fossil. You can see dinosaur fossils in Dinosaur Ridge, Dinosaur National Monument and many other places in Colorado.

There are even fossils of dino tracks and dino poop! A fossil poop is called a *coprolite*."
"Fossil poop? Really, Dad!"

minerals

fossils

dinosaur trackway

Pikes Peak and Garden of the Gods

"We know what this is!" shouted Emmaline and Oliver.
"Alright, let's hear it!" Dad said.

"It's a parade of pink and orange land sharks!" Mom laughed, "You two have some wild imaginations!"
"What is it really?" Oliver asked.

"If you came here hundreds of millions of years ago, you would have seen a different range of mountains, the Ancestral Rockies. Over time those mountains were destroyed by *erosion*.

Erosion is the breaking up of rocks and dirt by wind, water or ice. Bit by bit, grain by grain, water and wind carried the mountains away and those sediments washed down here to create layers of sedimentary rock such as sandstone.

Ancestral Rockies

Theiophytalia kerri

sediments

When Pikes Peak and the new Rocky Mountains were uplifted, the sedimentary layers were tilted. You can see tilted rocks like these in other places in Colorado, like the Flatirons in Boulder. Erosion is still shaping these rocks today."

GLOSSARY

coprolite Fossilized poop.
crust The earth's surface that we live on.
erosion The breaking up of rocks and dirt by wind, water or ice.
fossil A plant or animal, or something it left behind that has been preserved in rock.
glacier A large body of ice that forms on land and moves very slowly downhill.
lithosphere Greek for "rocky sphere." The rocky shell of a planet.
mantle A hot layer of the earth made up of partly melted rock and metal.
moraine A long deposit of gravel, cobbles and boulders pushed into place by a glacier.
plate tectonics A scientific theory that the lithosphere is divided into plates that move across the mantle.
sediment Fragments of rock like gravel, sand and silt that are moved around by water and wind.
subduction When one tectonic plate slides under another.

LEARN MORE, BE ACTIVE & HAVE FUN!

The best way to learn more about Colorado is to go there, or if you live there, to get out and explore!

Colorado is home to many amazing national and state parks, and some cities have their own great public lands like Boulder's Chataqua Park or Colorado Springs' Garden of the Gods. Don't forget to stop by museums and park visitor centers, attend ranger programs and become a Junior Ranger! You might even take a tour of an historic gold mine and learn to pan gold yourself. If you'd like to collect rocks or fossils, visit a BLM field office for guidance.

Stay active by hiking, biking, river tubing, sandboarding (in Great Sand Dunes N.P.), geocaching, skiing, snowboarding, sledding, or just playing in the woods or desert.

If you can't visit Colorado, visit nps.gov/webrangers. You can find some great games and videos on Colorado's national parks, build your own virtual ranger station and even earn rewards. Also try looking for books about Colorado at your local library.

Teachers can also find great resources, including curriculum, professional development and distance learning at nps.gov.

LANGUAGE ARTS EDUCATOR RESOURCE PAGE

MYTHS AND TALL TALES

DISCUSS: In the story, the kids and their parents give very different explanations for the natural wonders they visit in Colorado. Whose explanations do you like best? Which do you think are correct? Why? How do you think the parents and kids each came up with their explanations?

MYTH	TALL TALE
A sacred story that explains how people or things in nature became the way they are. Myths are often very important in religions or cultures and they are believed by followers.	A story that is unbelievable but told as if it is true. The story is told for fun. It might have started out as a true story, but then the details were exaggerated to make it more exciting or funny.

GO TO THE LIBRARY: Have students research and check out myths and tall tales. Read as a class, during silent reading time and at home.

ASK: As homework, have students gather myths or tall tales from parents, grandparents or others from the community. Back in class, they can share their stories and discuss whether they are myths or tall tales.

THINK: Ask each student to think of a natural feature or phenomenon and then imagine an extraordinary explanation for it.

WRITE: Ask students to record their stories in writing.

SHARE & DISCUSS: Read each others' stories aloud in class or in pairs. Discuss whether these are myths or tall tales.

EXTEND: Ask students to illustrate their stories and publish them as a book or rewrite them as scripts, make puppets and perform as a play.

© 2014 Eternal Summers Press. This page may be reproduced by educators for classroom use only. eternalsummerspress.com

SCIENCE EDUCATOR RESOURCE PAGE

THE ROCK CYCLE

PREPARE: You will need several examples each of sedimentary, metamorphic and igneous rocks. You may want to label them to avoid confusion. Gather information and materials for the demonstrations. Copy the chart below on a large piece of butcher paper.

THREE MAIN KINDS OF ROCKS		
SEDIMENTARY	METAMORPHIC	IGNEOUS

DISCUSS: What do we know about sedimentary rocks from *Imagination Vacation Colorado*? You might flip back to the pages on the formation of Garden of the Gods. Record responses in the chart. Add "MADE OF SEDIMENTS" to the Sedimentary column as well. Ask, "Does the word metamorphic sound like another word you know?" Talk about what metamorphosis means in relation to animals and ask what it might mean in relation to rocks. Record responses on chart. Add "CHANGED" to the Metamorphic column. Under Igneous, write "ignis = fire in Latin."

OBSERVE: Display the rock examples in front of their categories and allow students time to handle, make observations and record those observations on the butcher paper chart.

SEDIMENTARY DEMONSTRATION: Pour a 1/2 cup of fine sand into a "mountain" on one side of a plastic plate or tray. Explain to students that this represents the land. Mix a tsp. white glue into 1/3 cup water and as students watch, pour over the sand, allowing it to wash the sand across the plate. Ask, "what happened to the land?" Explain that this is similar to the erosion of land by water. Set the plate in a protected dry place. When the water has evaporated, show students the "sedimentary" rocks that have formed. Add "formed by weathering/erosion" to the chart under "sedimentary."

METAMORPHIC DEMONSTRATION: Pick your favorite cookie recipe (sugar or chocolate chip are great choices). Have students observe the characteristics of the ingredients before they have been blended and baked. Explain that these represent different kinds of rocks. Follow the recipe as usual but explain that during the mixing, kneading or forming of the cookies, you are applying *pressure* to the rocks. Before baking, ask students if the rocks look the same as in the beginning. Explain that you will now use heat to change the rocks even more. Bake the cookies as directed in the recipe. When the cookies are finished baking and have cooled, give one to each student and ask again how the rocks have changed. Can they even see the original rocks? Add "formed by heat and pressure" to the chart under "metamorphic." Explain that pressure can build up where tectonic plates run into each other. Heat is applied when magma (molten rock) from the mantle rises into the crust.

© 2014 Eternal Summers Press. This page may be reproduced by educators for classroom use only. eternalsummerspress.com

THE ROCK CYCLE (continued)

IGNEOUS DEMONSTRATION: For this demo, you will be making rock candy. There are a few ways to do it so select one (possibly from the internet) that you find user-friendly. Note: making rock candy involves boiling a solution; plan on having the students watch from a distance until everything has cooled. As in the other demos, explain that the sugar represents rock and have students observe the granulated sugar "rock" before the demo begins. Show students the pages that describe the formation of the Rockies and layers of the earth. Explain that the heat from the stove represents the heat in the earth's interior. Point out that it has melted the sugar, just like rock in a subduction zone is melted when it sinks into the mantle. It may take up to a week to grow sizeable candy crystals so set aside a place where students can easily observe the crystals growing without moving them. When you feel the crystals are large enough, have students once again observe the changes that have occurred. Compare the rock candy to a lollipop or hard candy and explain that igneous rocks can form with or without crystallization. Add "formed by melting and cooling" to the chart under "igneous."

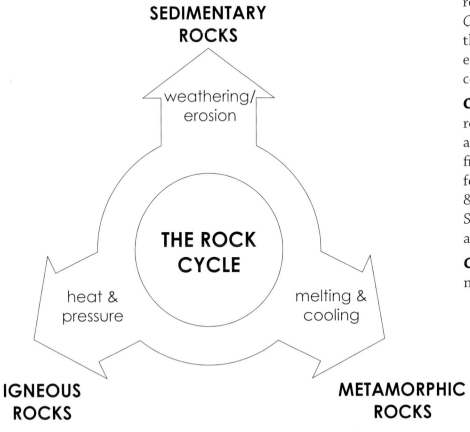

REREAD: As a class or in pairs, reread *Imagination Vacation Colorado* and discuss where the rock cycle is evident. Add examples to the appropriate column of the chart.

COMPLETE: Enlarge or draw the rock cycle diagram, leaving the arrows blank. Work as a class to fill in the arrows representing the forces (weathering/erosion, heat & pressure, melting & cooling). Students can add illustrations around the chart.

CELEBRATE: Your students are now Rock Stars!

VISUAL ART EDUCATOR RESOURCE PAGE
MAKE A FOSSIL

GATHER: Paper or Styrofoam bowls, non-hardening clay, a container for plaster mixing, drinking straws, watercolor paint in earth tones, twine, small plastic dinosaurs, insects, fresh real leaves or plants with heavy texture (choose sturdy ones) and sea shells. Read the entire lesson to get an understanding of the size/quantity of materials you will need. Keep in mind that the objects to be fossilized need to fit inside the bowl. Safety Note: Powdered plaster poses a respiratory risk and shouldn't be handled by students. Read all safety precautions for plaster.

INTRODUCE: Read the pages in *Imagination Vacation Colorado* that explain how fossils are formed. Allow students to handle some real fossils if possible or show photos of fossils (nps.gov/dino has great photos from Dinosaur Nat. Monument). Explain that in this activity the clay represents the ground and the plaster represents sediments.

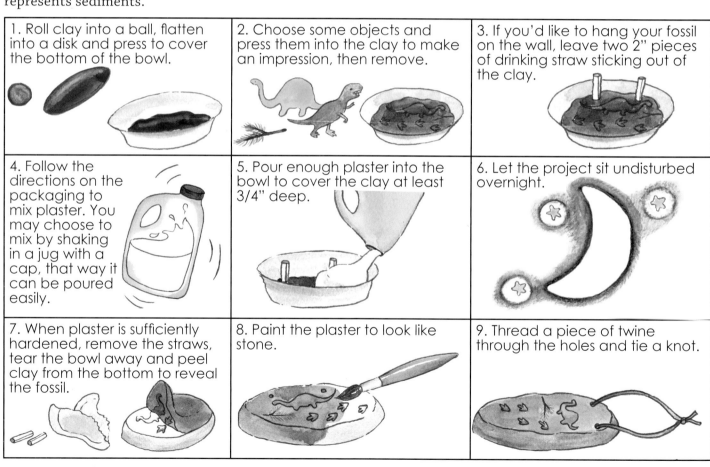

1. Roll clay into a ball, flatten into a disk and press to cover the bottom of the bowl.

2. Choose some objects and press them into the clay to make an impression, then remove.

3. If you'd like to hang your fossil on the wall, leave two 2" pieces of drinking straw sticking out of the clay.

4. Follow the directions on the packaging to mix plaster. You may choose to mix by shaking in a jug with a cap, that way it can be poured easily.

5. Pour enough plaster into the bowl to cover the clay at least 3/4" deep.

6. Let the project sit undisturbed overnight.

7. When plaster is sufficiently hardened, remove the straws, tear the bowl away and peel clay from the bottom to reveal the fossil.

8. Paint the plaster to look like stone.

9. Thread a piece of twine through the holes and tie a knot.

REFLECT: Ask students how this project is different from real fossil formation.

© 2014 Eternal Summers Press. This page may be reproduced by educators for classroom use only. eternalsummerspress.com